Dinosaurs

10 Things You
Should Know

Dr Dean Lomax is an internationally recognised, multi-award-winning palaeontologist, author and science communicator. He travels across the world, excavating and researching dinosaurs, discovering new species, and regularly appears on TV as an expert and presenter, notably co-hosting the prime-time TV series *Dinosaur Britain*. He has written several books and many academic papers, is a leading world authority on ichthyosaurs and has given a TED Talk on his unusual path to becoming a palaeontologist. He won a gold medal for excellence in science at the Houses of Parliament in 2015 and was one of twenty finalists for the 2018 JCI Ten Outstanding Young Persons of the World award – an award won previously by the likes of Elvis Presley, Jackie Chan and Bill Clinton. He is a patron of both the UK Association of Fossil Hunters and Mary Anning Rocks. Follow him on social media @dean_r_lomax and visit his website at: www.deanrlomax.co.uk.

Dinosaurs

10 Things You Should Know

230 million years for people short on time

DR DEAN LOMAX

SEVEN DIALS

First published in Great Britain in 2021 by Seven Dials
an imprint of The Orion Publishing Group Ltd
Carmelite House, 50 Victoria Embankment
London EC4Y 0DZ

An Hachette UK Company

5 7 9 10 8 6

A CIP catalogue record for this book is
available from the British Library.

ISBN (Hardback) 978 1 8418 8494 3
ISBN (eBook) 978 1 8418 8495 0

Printed in Great Britain by Clays Ltd, Elcograf S.p.A.

FSC
www.fsc.org

MIX
Paper from
responsible sources
FSC® C104740

www.orionbooks.co.uk

To Elaine Howard, who followed her passion
for dinosaurs to the ends of the Earth.
And to you, reader, for continuing to pursue
your passion for the past.

Oh, and to my family, for trying to show
me that there is more to life than playing
with dinosaurs. There isn't.

Contents

Timeline

CENOZOIC ERA	Quaternary	2.6 to present day
	Neogene	23 to 2.6
	Palaeogene	66 to 23
MESOZOIC ERA	Cretaceous	145 to 66
	Jurassic	201 to 145
	Triassic	252 to 201
PALAEOZOIC ERA	Permian	299 to 252
	Carboniferous	359 to 299
	Devonian	419 to 359
	Silurian	444 to 419
	Ordovician	485 to 444
	Cambrian	541 million to 485

Precambrian   4.5 billion to 541 million

Preface

Dinosaurs are the ultimate symbol of prehistoric life. These awesome ancient creatures supercharge our inquisitive minds and help us to appreciate and understand that the world is not everything we see around us today. The fossilised remains of dinosaurs and other long-lost creatures from the depths of time tell us that life on Earth is temporary. That we are all part of one gigantic story – the evolution of life.

We are truly mesmerised by dinosaurs, especially as children. There is nothing more captivating than seeing a long-necked giant the length of three school buses or an enormous predator with massive bone-crunching teeth unlike anything alive today. In many ways, dinosaurs seem so fantastically impossible that they can be characterised as having a superhero-like status. And yet, they are real. We learn about dinosaurs through books, toys, movies, TV and, above all, museums where we can

1

actually see their skeletons and occasionally even touch their bones. For most people, dinosaurs are their first introduction to science, a unique gateway of discovery that allows young and curious minds to begin asking 'why?' 'Why did the dinosaurs live so long ago? Why did they go extinct? Why were some of them so big?' Although many of us grow out of the so-called 'dinosaur phase', we never stop appreciating these extraordinary animals, keeping our passion for the past very much in the present.

Nowadays it seems you cannot go anywhere without seeing a dinosaur. I'll be walking down the frozen-food section in the supermarket and there it is, a friendly-faced dinosaur taunting me to buy that box of ice lollies. Looking at birthday cards and ten different dinosaur-themed versions are staring back at me, usually depicting some big toothy-grinned dinosaur or a joke about the short arms of *Tyrannosaurus*. Dinosaurs are as hip as they have ever been, featuring even as mascots for sports teams, in fashion jewellery, designer clothing, household ornaments, on TV adverts, and much more. Our fascination with everything dinosaur will never die.

Admittedly, I might be a little bit more conscious of all these dinosaur titbits because I am one of those people who was fascinated by dinosaurs as a child and never grew up. Today, I live out my childhood dream of being a palaeontologist, a scientist who studies dinosaurs, fossils and the evolution of life. Palaeontologists often get confused for archaeologists, people who study human history and prehistory, and occasionally get labelled as 'Indiana Jones'. Well, either that or, 'Oh, like Ross from *Friends?*', or the inevitable *Jurassic Park* reference. It's okay if you were thinking it, we're all used to it by now.

For me, there is nothing quite like that mind-blowing feeling of being the first person *ever* to unearth a multimillion-year-old dinosaur skeleton from its rocky tomb or realising that you have found a new species that you get to name. For that brief moment in time, you are the only person in the entire world who knows this. The incredible experience of discovery is, of course, not just restricted to palaeontologists because anybody can discover a fossil – you just have to know where to look.

You can imagine each dinosaur discovery as being a very tiny piece of an enormously complex jigsaw puzzle with the cover image long gone. Palaeontologists have to work with all of these minute dinosaur pieces to help build up a bigger picture of the ancient dinosaur world; our understanding of which has changed dramatically over the past thirty years. With new discoveries come new knowledge, such is the wonderful ever-changing nature of science.

Combining the rich history of dinosaur discovery with the latest research, in this book I have carefully pieced together ten chapters which provide you with the basic essentials of everything you need to know about dinosaurs. By taking this grand tour many millions of years into the past, you will discover when and where dinosaurs lived, how they copulated and became populated, and when their epic reign at the top came crashing down from the heavens. This book will give you an exciting glimpse into the science of dinosaurs, bringing to life an ancient world that continues to evolve with new discoveries, forever expanding our knowledge of the most magnificent animals ever to have walked the Earth.

1

Why *Tyrannosaurus rex* and *Stegosaurus* Never Met

Earth is estimated to be a staggering 4.54 billion years old – a rather unimaginable amount of time. For contrast, at the time of writing this book, 1 billion seconds ago was the late 1980s or 31.7 years ago; I guess that makes me almost 1 billion seconds old. To help place this grand scale of time into context, just for a moment imagine cramming all of geologic time into one calendar year, with 1 January marking the birth of the Earth. Dinosaurs would appear on 13 December and modern-day humans around ten minutes before midnight on New Year's Eve.

In order to make sense of the vast age of the Earth, the geologic timescale is divided into many distinct chunks of time based on important changes in the geologic record, such as an extinction event or the appearance of new species. Time is split into eons, eras, periods, epochs and ages (or stages), along with numerous smaller subdivisions, whose boundaries are precisely defined in age through a technique called *radiometric dating*, which we'll discuss a little further on. This is an ever-evolving system studied by geochronologists, scientists who specialise in dating geologic ages. As new information is obtained, the geologic timescale continues to be revised, updated and improved, which means the age of the boundaries sometimes changes.

There are four major chapters of time, known as eons, which comprise, from oldest to youngest, the Hadean, Archaean, Proterozoic and Phanerozoic. The first three are collectively referred to as the Precambrian and account for a staggering 88 per cent of geological time (the first 4 billion years), from the formation of the Earth up until the beginning of the Phanerozoic, 541 million years ago (or 18 November on the one-year calendar). We know that life originated during the Precambrian because the oldest fossil evidence of simple, single-celled microscopic organisms is recorded from rocks that are at least 3.5 billion years old. Fast-forward towards the end of the Precambrian, about 570 million years ago, and here is where the first evidence for complex multicellular life emerged of strange soft-bodied creatures that made their lives on the seabed. Even at this point in geologic time, the story of the dinosaurs remains more than 300 million years away.

Why *Tyrannosaurus rex* and *Stegosaurus* Never Met

The Phanerozoic is the eon that we are currently in, where life evolved and diversified on an enormous scale, and where fossils are abundant. Extending from 541 million years ago until the present day, the Phanerozoic Eon is divided into three eras, the Palaeozoic (meaning 'ancient life'), Mesozoic (meaning 'middle life') and Cenozoic (meaning 'new life'). During the Palaeozoic, life exploded onto the scene, animals with hard shells appeared, the first vertebrates evolved and eventually took their steps on land, and the era ended with the greatest mass extinction event ever known, wiping out as much as 90 per cent of all life. Rising up from the ashes of the extinction, the Mesozoic witnessed the evolution of the dinosaurs, the emergence of mammals, and the first flowering plants. Finally, in the Cenozoic, the very same era of geologic time that we are in today, mammals dominated the world and things appeared more similar to how they are today.

In the grand scale of Earth's vast past, when plotting the dinosaurs onto a geologic timescale like this, it becomes incredibly apparent that they appeared only very recently. It is even more amazing to consider that there were already fossils under the feet of dinosaurs, of countless species that had come and gone, long, long before the dinosaurs had even evolved.

The Mesozoic Era is often labelled as 'the age of dinosaurs', when these ruling reptiles ran the world. For many, this dinosaur world is imagined as one where all the dinosaurs lived together at the same time. This vision comes from the influence of movies, TV shows and sometimes books, especially those that we watched or read as children and which became ingrained in our dino-crazed minds. Take, for example, the popular *The Land Before Time* franchise, which is a personal favourite of mine, where all types of different dinosaurs and other prehistoric animals are portrayed as living together at the same time. Yet, dinosaurs did not all live at one set time in the Mesozoic Era. In reality, this slice of dinosaur life is split into three periods – from the oldest to the youngest, the Triassic, Jurassic and Cretaceous – and extended a whopping 186 million years, from 252 to 66 million years ago.

Furthermore, although the Mesozoic might be considered 'the age of dinosaurs' on the whole, the dinosaurs did not live for the entire length of the Mesozoic and were not always the 'rulers' either.

See, the very earliest known dinosaur fossils were found in the badlands of Ischigualasto (that's fun to say!) in Argentina, South America. They show that dinosaurs appeared during the latter part of the Triassic Period, from a little over 230 million years ago. That means the Earth's oldest known fossils (from 3.5 billion years ago) are fifteen times the age of the earliest dinosaurs. Many of these early dinosaurs were fairly modest in size and were overshadowed by their contemporaries, like the predatory crocodile-like animals called phytosaurs and car-sized carnivorous amphibians, which very likely dined on these early dinosaurs.

You might be wondering how we can possibly know when the oldest dinosaurs were alive, or how we can determine when any dinosaur lived for that matter, especially as no dinosaur was buried with its death certificate. To begin to understand this, we have to go back to the 1600s and to a Danish scientist called Nicholas Steno, who came up with a very simple way of interpreting the relative age of rocks (and the fossils therein).

In 1669, Steno proposed what became known as the 'principle of superposition', which is today one of the fundamental principles of geology. Simply put, in any undisturbed sequence of rocks deposited in layers (called strata), the oldest rocks are on the bottom and the youngest are on the top. An exceptional example can be observed in the breathtaking and richly diverse geology of the Grand Canyon, where the most ancient rocks are at the very bottom and each layer above gets progressively younger the closer you get to the top. You can think about it like the layers of a cake, the bottom layer being laid down first and the top layer laid down last. So, as a basic rule, in the most perfect undisturbed circumstances, this means that fossils from deeper layers are geologically older than fossils from those layers above.

Today, Steno's work acts as the starting point for understanding the relative age of rocks, but a new technique developed in the early twentieth century was a total game-changer. Called radiometric dating, this technique enabled geologists to provide solid estimates for the absolute age of certain rocks by carefully studying the steady rate of decay of radioactive elements contained inside them. Using a combination of both methods, geologists and palaeontologists can determine the age of rocks and thus correctly identify which of the three periods and exactly where in those periods a dinosaur bone came from. This process is similar to, and is often confused with, the more familiar radiocarbon dating, or carbon-14 dating, where scientists study the decay of radioactive carbon-14 atoms to determine the age of an object. However, radiocarbon dating could never be used to date a multimillion-year-old dinosaur bone because radiocarbon decays rapidly (geologically speaking) and is only used for dating anything in the relatively recent geological past, within the last ~50–60,000 years or so.

The Triassic, Jurassic and Cretaceous periods differ substantially in the length of geologic time, and to better understand the differences between these three periods it is best to look at how dinosaurs are placed within them. Dinosaurs that lived during the same time period did not necessarily live together at the same time and could have lived at various intervals within that period – even being separated by millions of years. Typically, any given species only lived for a couple of million years before it went extinct. In addition to this, those dinosaurs that did live together at the very same time may have lived on a different continent entirely.

By studying the rocks in which a dinosaur fossil is entombed, therefore, palaeontologists have been able to deduce with confidence that a *Stegosaurus* lived during the Jurassic Period and a *Tyrannosaurus* lived during the Cretaceous Period. It seems simple enough to say that *Stegosaurus* and *Tyrannosaurus* never met because they lived in entirely different periods of time. However, there is a lot more to it than that. What exactly does it mean to be from the Jurassic versus the Cretaceous?

Let's first look at *Stegosaurus*, my childhood favourite dinosaur. When I refer to a dinosaur's age, as in '*Stegosaurus* lived during the Jurassic Period', there is a specific time interval within the Jurassic that I have in mind. That is because the Jurassic Period began 201 million years ago, ended 145 million years ago, and is split into three epochs: the Early, Middle and Late Jurassic, which are divided into eleven different stages. Fossils of *Stegosaurus* are from the Late Jurassic, a chunk of time that extended from 163.5 to 145 million years ago. More specifically, the fossil remains come from the last two stages in the Late Jurassic, called the Kimmeridgian and Tithonian, which extended from 157 to 145 million years ago, although *Stegosaurus* fossils are known only from rocks that are approximately 153 to 148 million years old. (I know that might seem quite complicated but stay with me.)

Now, compared with *Tyrannosaurus*, which lived towards the very end of the Late Cretaceous, between 68 and 66 million years ago in a stage known as the Maastrichtian, by doing the maths you can see that *Stegosaurus* was already extinct 80 million years *before Tyrannosaurus* even walked on the Earth! Even more impressive, the oldest tyrannosaur fossils, aka grandparent rex, are more than 160 million years old. To really hit home the magnitude of geological time, *Tyrannosaurus* is closer to you and me in time than it was to *Stegosaurus* or even great-grandparent rex. Let that sink in!

2

Globetrotters

Just for a brief moment, pause and think about the places you have travelled to in your lifetime. Have you ventured to the far-flung corners of the world, ticked off many locations on your bucket list, and stepped foot on different continents? Despite the surface of the Earth being covered up to 70 per cent by water, having the luxury of flying on planes, boarding a boat and taking the train we can pretty much travel anywhere we want in the world. No ocean is gonna stop us. The dinosaurs, on the other hand, did not have access to boats or planes, but that did not stop them from travelling across the world either. Not because they were Olympic swimmers, but because their early world was stuck together in one giant supercontinent.

In our schooldays we are taught that there are seven major continents in the world – from largest to smallest, Asia, Africa, North America, South America, Antarctica, Europe and Australia. Smaller land masses, like many of the world's big and small islands, are often grouped together with one of the neighbouring continents, and so, the continent of Australia is grouped with islands in the Pacific

Ocean to form the region called Oceania. We are also taught that these continents are still on the move, albeit very slowly, at speeds of just a few centimetres each year. The continents rest on enormous chunks of the Earth's crust called tectonic plates, which move over liquid rock located deep beneath them. When they move, these tectonic plates may pull apart, collide together, or slide past each other, effectively changing the location of the continents and reshaping the face of the Earth.

The evidence contained inside rocks dating from millions and billions of years ago shows how mountain ranges have risen and crumbled, that epic seas have come and gone, and how continents have crashed together and drifted apart. For example, have you ever looked at a map of the world and noticed that the eastern coast of South America appears as if it could fit into the western coast of Africa, a bit like a ginormous jigsaw puzzle? This is no coincidence. The continents have not always been where they are today, and neither has the place where you are reading this book right now. In fact, this is just one piece of the overwhelming evidence that suggests all of the Earth's continents must have been joined together in a single, enormous land mass that eventually separated. And not just once. Rather, Earth's continents have been combined as a supercontinent on multiple occasions.

A supercontinent is defined as an enormous single continent made up of all, or nearly all, of the Earth's land. The first supercontinents formed more than a billion years ago. They lasted for a few hundred million years before breaking apart and eventually coming back together again hundreds of millions of years later, where the process started over again. This is called the supercontinent cycle. Scientists have even predicted that the next supercontinent will form within the next 200–250 million years. However, we are interested in the most recent giant supercontinent, called Pangaea, which played a crucial role in helping to shape the story of the dinosaurs and give them a fighting chance at success.

The Earth was a very different place when the dinosaurs appeared on the scene in the Triassic Period. With all of the world's continents united together, positioned in such a way that northern South America, Africa, southern North America and Europe were equatorial, this alien world looked nothing like it does today. If you happened to live in what is now Britain during the Triassic Period, you would have been much closer to the equator and lived in a scorching hot and dry desert.

Stretching almost from pole to pole and shaped sort of like a giant thick 'C', Pangaea's name which comes from Ancient Greek and means 'Whole Earth', really was an enormous land mass. It was surrounded by a global super ocean named Panthalassa, which was the ancestor of today's Pacific Ocean. The supercontinent actually formed around 300 million years ago, about 70 million years before the dinosaurs appeared, and provided the necessary platform for global dinosaur domination – eventually.

Living on a giant supercontinent with no oceans to cross meant that, in theory, any animal or plant could spread right across the world. However, it is not that simple and no one species spread throughout all of Pangaea. Animals were restricted from doing so due to things like extreme weather or challenging terrain, although evidence does show that some species did travel far and wide across parts of the supercontinent. One of the interesting findings that helped to support the idea of Pangaea was the discovery of the same types of fossil animals and plants from identical types of rock found on different continents in the southern hemisphere. Some of these animals were clearly adapted for a life on land and it would have been impossible for them to swim across oceans that separate the continents today.

By the end of the Triassic Period, dinosaurs had already spread across various parts of Pangaea. When the mighty supercontinent began to break up during the Early Jurassic, about 200–180 million years ago, it eventually split into two enormous land masses, Laurasia to the north and Gondwana to the south. The land became separated by vast oceans and drifted further apart, where dinosaurs became isolated across the globe on different continents. This meant that dinosaurs in the north were subjected to different environments from those in the south, which led to the evolution of very different species that were better adapted to thrive in their new worlds. Over millions of years, the continents continued to split, changing shape and position and eventually the continents as we know them today began to take shape.

This is how dinosaurs therefore conquered the world. Their remains have been unearthed on every single continent, collected from high mountain ranges and deep underground mines. But beyond being able to say that dinosaur X was found on one continent and dinosaur Y was found on another, perhaps the more intriguing thing about discovering dinosaurs (or any other fossil) is revealing the story of their world contained within the rocks that entombed them.

Take a moment to imagine you are part of an epic fossil-hunting team trekking deep into the heart of Africa's Sahara Desert, when you come across fossils of animals that once lived in the sea, such as corals. You know that corals are found in the sea today but there is no sea around for miles and miles. What the fossil tells you is that the area in which you are standing was once underwater millions of years ago when the creature was alive. For a palaeontologist, having evidence of past environments and ecosystems like this provides a more complete picture of the animal and the world in which it lived, and, in some cases, these details can be even more important than the fossils themselves.

Using this same approach for dinosaur discoveries, one of the most fascinating and unexpected examples is that several dinosaurs have been found in Antarctica, contained inside rocks that have long since been buried underneath the ice. Better still, an abundance of different fossils has been found alongside them, including numerous species of plants. What these discoveries have revealed is that the ancient environment changed throughout geologic time and was dramatically different from the extreme icy conditions of Antarctica today – during the reign of the dinosaurs, at certain times, it was even a warm and lush rainforest.

Looking at another remarkable discovery, about 70 million years ago in the Cretaceous Period there once existed an island located in what is now Romania, called Haţeg Island. The island was cut off from the rest of the world and was ruled by dwarf dinosaurs – an example of insular dwarfism, where larger animals shrink over generations to adjust to a smaller environment that has less food and fewer predators. These types of incredible discoveries allow us therefore to understand how the Earth as we know it has changed so dramatically over the course of its very ancient history and how the mighty dinosaurs were forced to adapt in order to dominate all corners of their world.

3

What Makes a Dinosaur
a Dinosaur?

One of the questions I get asked all the time is: 'What makes a dinosaur a dinosaur?' You might think this should be a pretty straightforward answer for a palaeontologist, but it's surprisingly hard. In the mind of a palaeontologist, the word dinosaur means something very specific. And, almost like ticking off a dino checklist, in order to be called a dinosaur, a combination of distinguishing features must be preserved.

The first rule of dino club is that the word dinosaur is not a catchall name for anything that is extinct. Neither does it mean any old prehistoric reptile. To look at this another way, consider the amazing diversity of mammals today. You know that a mouse is a mammal, as are you and I, but also that a blue whale is a mammal, and so is a bat. Yet, these mammals all belong to different groups and families. Just because dinosaurs are reptiles does not mean that all prehistoric reptiles are dinosaurs.

The fossilised remains of dinosaurs and other ancient creatures must surely have been discovered throughout the entire history of humankind. To understand where it all began, we have to travel back in time a little bit, not to the Jurassic, but to 1824. This is a particularly significant year because it is when the very first dinosaur (which is actually from the Jurassic) was formally recognised by science and officially announced to the world.

Lots of fragmentary bones, and one particularly impressive-looking jaw with large serrated teeth, were discovered inside underground slate mines near the village of Stonesfield, in Oxfordshire. These bones eventually wound up in the hands of one of the greatest geologists of the time, William Buckland, a reverend and professor of geology at Oxford University. After studying the bones, in 1824 Buckland announced the discovery of a mysterious giant lizard-like reptile which he named *Megalosaurus*, or the 'great lizard'. Just a year later and another of these 'giant lizards' was revealed, called *Iguanodon*, this time named by a country doctor-turned-geologist, Gideon Mantell. Mantell went on to describe another of these strange beasts in 1833, which he called *Hylaeosaurus*. At the time of their discovery, nobody knew exactly what kind of animals they were, but all that was about to change.

More bones and teeth of these curious creatures popped up across the British countryside and astounded scientists and the public alike. The original trio of animals, along with additional fossils, were examined by leading comparative anatomist and palaeontologist, Sir Richard Owen, who was the founder of the Natural History Museum in London. Owen noticed similarities in the hip bones, particularly the presence of five fused vertebrae (called the sacrum), and how the limbs were held under the body. He realised that they belonged to a unique group of animals for which he coined the word 'Dinosauria' in 1842, taken from the Greek words *deinos*, meaning 'terrible' or 'fearfully great', and *sauros*, meaning 'lizard'.

When you look at the discovery of dinosaurs in this timeframe, it is quite amusing to think that, despite them appearing more than 200 million years ago and leaving their mark on our planet in such a dramatic way, dinosaurs are a very recent 'invention'. Not even two centuries old. The first steam engines were running on rails before the dinosaurs received their name.

In the vast majority of cases, multimillion-year-old dinosaur fossils are represented by isolated bones or fragmentary skeletons rather than complete individuals. This is one of the major challenges that we palaeontologists have to face when studying, describing and defining what a dinosaur is. That being said, one of the amazing things about science is that it is in a constant flux. The very concept of what actually constitutes a dinosaur has come a long way since Owen's 'terrible lizards' and is still tweaked when major new discoveries are made.

Without getting into all of the nitty-gritty finer details of bony features in the skull and skeleton, which palaeontologists fuss over and cannot quite agree upon, there is a combination of particularly significant features that are used to *quickly* identify a dinosaur.

Firstly, the next time you are in a museum staring at a skeleton of some extinct *-saur*, quickly glance at the head and look at the number of holes in the skull. To be a dinosaur you must first be a *diapsid*, aka a reptile with two holes (openings) on either side of the skull, behind the eye socket. Within the diapsids, dinosaurs belong to a group called archosaurs, or 'ruling reptiles', which today include the crocodylians and birds. All archosaurs, including the dinosaurs, then have an additional hole in the skull situated between the eye socket and the nostril called the antorbital fenestra.

Then turn your attention to the hips. Just as Owen had originally highlighted, the hips play a critical role in the modern dinosaur definition. Notably, the presence of a distinct hole in the hip socket, formed between the three hip bones, is especially important and marks the place where the 'ball' or head of the thighbone (femur) locked into position. This allowed dinosaurs to adopt an upright stance like mammals, with their limbs held directly beneath their body, unlike crocodiles and lizards whose limbs are sprawled out to the side. This is the primary feature that helps to set dinosaurs apart from other reptiles.

Taking a step back from the bones, based on the evidence we have of countless dinosaur eggs, palaeontologists are confident that egg-laying was ubiquitous among dinosaurs. Furthermore, although some species definitely spent time in the water feeding or swimming, dinosaurs were fully adapted to a life on land and no dinosaur lived exclusively in the watery realms.

The hip bones also play a central role in how the dinosaur family tree is split. Traditionally, based on the shape and position of the hips, the dinosaurs are divided into two specific branches, called the

Saurischia and the Ornithischia. The saurischian dinosaurs have 'lizard-like hips' and include dinosaur hotshots like *Tyrannosaurus* and *Diplodocus*. Whereas the ornithischians have 'bird-like hips' and include such dinosaur celebs as *Stegosaurus*, *Triceratops* and *Iguanodon*. Each of the dinosaurs is then placed into a distinct group or family within either the Saurischia or Ornithischia based on more specific features.

In 2017, one team of palaeontologists published a significant study that proposed a major shake-up of the overall shape of the family tree, suggesting that the traditional split was not quite correct. They argued that dinosaurs like *Tyrannosaurus* are more closely related to the likes of *Iguanodon* and *Triceratops* than *Diplodocus*. However, palaeontologists have yet to wholly agree whether this new divide is more accurate than the previous, which has been in place since 1888. This hypothesis will continue to be tested over the coming years and new specimens and discoveries will no doubt further our understanding.

Although palaeontologists do not always concur on the finer points, one thing that we all agree upon is what does *not* make a dinosaur. By some distance, there are two main groups of animals that are frequently labelled as dinosaurs but are not. First up, the 'flying dinosaurs', aka the pterosaurs or, if you must, pterodactyls. (Pterodactyl is not quite correct because this name refers to a particular pterosaur called *Pterodactylus*, the very first pterosaur to be discovered.) Pterosaurs were a wondrous group of flying reptiles that lived during the Mesozoic Era. Long before birds or bats, they were the first vertebrate animals to evolve flight and had wings that stretched from the tips of their fingers to their ankles. Like the dinosaurs, pterosaurs are also archosaurs, but differ greatly in their anatomy and are on another branch of the archosaur tree of life. You can think of them as dinosaur cousins.

Next are the so-called 'swimming dinosaurs'. It hurts me to say that, because they are my favourites. I am of course speaking about marine reptiles, which I could write a book on (perhaps that should be next on my agenda). These include the dolphin-like ichthyosaurs, mighty lizard mosasaurs and, please do not make me say it, 'Loch Ness Monster-inspiring' plesiosaurs. Well, you made me say it.

These animals are much further removed from

the dinosaurs, in evolutionary terms, and lived exclusively in water, even giving birth to live young out at sea. One of the major differences from the dinosaurs is that their limbs were modified into fins, which they used to manoeuvre through the water as they swam with their tails. In particular, the ichthyosaurs have a special place in my heart because most of my academic research has been dedicated to them. I've spent years studying and measuring the bones and teeth of thousands of ichthyosaurs, naming five new species in the process and identifying one that was probably as big as a blue whale before blue whales were even a thing. In fact, long before whales were even a thing.

Finally – to the extreme – fake news, poor journalism and cheap dinosaur books would have you believe that practically anything prehistoric is a dinosaur. I have seen headlines stating that even a woolly mammoth is a dinosaur, simply because it is *old*. It makes me shudder every time. Similarly, have you ever opened a pack of dinosaur toys and discovered that lizard-like sail-backed 'dinosaur'? No, not *Spinosaurus*, but *Dimetrodon*? This dino imposter is pretty much the non-dinosaur dinosaur poster child of the prehistoric world. The funny thing is that *Dimetrodon* is not a dinosaur or even a reptile, but a protomammal that is more closely related to us than to reptiles. A distant relative, as it were. So, the next time you see a headline shouting about some new 'dinosaur discovery', be sure to go in armed with all of this newfound knowledge and decide for yourself whether this is a true dinosaur or just another wannabe.

4

Velociraptor Was the Size of a Turkey

Name five dinosaurs right now, go! *Velociraptor* immediately comes to mind, not simply because it is in the chapter title but because it is a contender for the most famous of all the dinosaurs, being featured in practically every dinosaur book as well as being a star on the big screen. When working with the public I like to ask people what their favourite dinosaur is, or to ask them to name five dinosaurs. This is not to put them on the spot (although maybe sometimes it is, as I like to see if they say 'pterodactyl' or ichthyosaur), but because I am genuinely interested to uncover which dinosaurs are fixed in people's minds. I am yet to meet anybody who has not included *Velociraptor* in their five dinosaurs.

When you tell people that *Velociraptor* was actually the size of a turkey but with a long tail, many look at you with a puzzled expression. You can see they are questioning your palaeontology credentials: 'Is this person really a palaeontologist? Next they'll be telling me that *Velociraptor* had feathers …' The issue is that, at least in pop culture, *Velociraptor* is visualised as being as tall as an adult human, although the science very clearly shows otherwise. You see, this idea of an oversized *Velociraptor* has its roots in *Jurassic Park*, yet these big-screen 'raptors' are not really *Velociraptor* at all. Instead, they were based on a close cousin, called *Deinonychus*.

When Michael Crichton was looking for would-be dinosaur stars to feature in his novel, *Jurassic Park* (1990), one palaeontologist had suggested in a popular dinosaur book that *Deinonychus* and *Velociraptor* were so similar that they should be grouped together under the name *Velociraptor*. This despite the fact that they lived at different times, were found continents apart and had numerous anatomical differences. Evidently, palaeontologists at the time did not agree with the name change.

Crichton, who was apparently inspired by that popular book, even acknowledging the author in his novel, now had a fleet-footed, sickle-clawed deadly *Velociraptor* approaching the size of a human rather than a turkey. To better understand how these dinosaurs lived and hunted, Crichton met with and quizzed the influential American palaeontologist Professor John Ostrom, the person who discovered *Deinonychus*, whom he also acknowledged in the novel. Interestingly, when making the movie, Spielberg's team actually studied Ostrom's scientific papers that described *Deinonychus*.

In the end, *Velociraptor* was the name that made the cut in the novel and subsequent film adaptation in 1993. Some accounts have suggested that Crichton merely thought that *Velociraptor* sounded more dynamic, catchier, and more menacing than *Deinonychus*, and this was his reason for choosing the name. Funnily enough, in one of the movie's opening scenes, a team of palaeontologists is shown excavating a *Velociraptor* skeleton in Montana, USA, although the fossils of this dinosaur have only ever been found in Asia. By comparison, *Deinonychus* is known only from the USA and the first discovery was made in Montana.

While *Velociraptor* is in the spotlight, I feel it is my duty as a palaeontologist to point out that it did not have the dexterity to open doors, let alone reach the handles. Unlike in the movie, it did not have the infamous 'bunny hands', where the palms of its hands face the ground. Rather, the palms faced each other (think about when you clap) and were perfectly positioned for slashing at prey with its three curved claws.

The size issues surrounding *Velociraptor* are interesting because for many people the word dinosaur is often synonymous with something big and intimidating. From this perspective, in the public's imagination I guess an overgrown *Velociraptor* sort of makes sense. One of the biggest and most common misconceptions about dinosaurs is that they were all large.

Ask any palaeontologist about dinosaur size and they will surely be quick to mention the biggest of the big, the sauropod dinosaurs. You know the types, those herbivorous behemoths with tremendously long necks and long tails and which include dino-stars like *Brontosaurus* and *Brachiosaurus*. It is mind-boggling to imagine, but reliable estimates have shown that some sauropods reached upwards of 30 metres (98 feet) in length and about 70 tonnes in weight – the equivalent average weight of twelve fully grown African elephants. Sauropods were true titans of the past and include the largest animals ever to walk on Earth. By contrast, today the largest terrestrial animals are elephants (for weight) and giraffes (for height).

As for the big meat-eaters, *Tyrannosaurus* is still one of the largest at about 12–13 metres (39–43 feet) long and 8 tonnes, although the bizarre sail-backed *Spinosaurus* currently takes the top spot for length, being a couple of metres longer still. By comparison, it is quite laughable to consider that the largest terrestrial carnivore today is a polar bear. Yes, a polar bear. Think about the size of a *Tyrannosaurus* compared with a polar bear, which typically weighs half a tonne and measures about 2.5 metres (8 feet) long.

Truly, these were enormous animals. The reality, however, is that dinosaurs varied immensely in size and many of them were tiny, at around the size of a squirrel. Palaeontologists have known this since the 1850s following the discovery of the carnivorous chicken-sized *Compsognathus*, hailed for decades as the smallest dinosaur in the world (just go and check some of your older dinosaur books) although there are many contenders today for the title of world's smallest extinct dinosaur.

Like *Compsognathus*, virtually all of these pint-sized dinosaurs are small meat-eating theropods, the same group of dinosaurs that includes such mega-predators as *Tyrannosaurus* and *Megalosaurus*. Given that many are roughly the same length, around the 30–50 centimetre (12–20 inch) mark, palaeontologists find it difficult to say for certain which was the king of the tiny terrors, although *Epidexipteryx* from China comes close. Weighing in at about 160 grams (6 ounces) and measuring approximately 25–30 centimetres (excluding its tail feathers), this Jurassic minisaur named in 2008 was about the size of a pigeon.

Over the past thirty years or more, Asia has become the epicentre for palaeontology discoveries. Many new finds have radically changed our understanding of dinosaurs, especially how they looked. Perhaps the major discovery that really announced China on the world stage for palaeontology came in 1996 with the discovery of the world's first feathered dinosaur. Depending on how old you are, you may remember hearing about the discovery in the news. I was only six years old at the time and so was probably busy playing with my featherless toy dinosaurs.

Named *Sinosauropteryx*, the fossils of this Cretaceous-aged cousin of *Compsognathus* were extraordinarily well preserved, containing a plumage of primitive feathers (or 'dino-fuzz') and soft tissues. It would eventually go on to become the first dinosaur to have its true colour scientifically established. Multiple studies of tiny colour pigments (called melanosomes) locked inside the feathers showed that it was orangey-red and white, with a striped tail and raccoon-like 'bandit mask' on its face. The colouration also showed that it was darker on top and lighter underneath, representing a form of camouflage known as countershading.

Staying with feathered dinosaurs, simply flick through old dinosaur books or search the internet for an image of our friend *Velociraptor* and you will find countless reconstructions of a lizard-like scaly dinosaur lacking any signs of feathers. We now know, thanks to fossils found in Mongolia, that *Velociraptor did* have feathers.

A neat study published in 2007 presented direct evidence for feathers in *Velociraptor* based on a row of evenly spaced bumpy 'quill knobs' on the forearm. Quill knobs are found in many living birds and identify the bony anchor points for big wing feathers. This is a clear indication that *Velociraptor* had wings, although its body size and relatively short forearms show that it definitely could not fly. Instead, the wings were very likely to have been used for display, perhaps to control its temperature and/or to help it manoeuvre while running. So, any time you see a reconstruction of *Velociraptor* lacking feathers, you can just imagine that somebody has plucked it ready for the oven.

And it is not just Asia where these types of revolutionary discoveries are being made. In 2017, news broke of an exceptionally well-preserved armoured dinosaur (an ankylosaur), named *Borealopelta*. Collected from inside a mine in Canada, this stunning dinosaur is three-dimensionally preserved and looks as if it simply went to sleep or was 'mummified'. There is evidence of preserved traces of tiny colour pigments trapped inside its fossilised skin, as well as keratinous horn sheaths covering its armour (keratin is the same material that makes up our nails and hair). Studies of the preserved colour showed that it was reddish-brown and also displayed countershading, similar to *Sinosauropteryx*.

Practically every year a major discovery is made in palaeontology that somehow revolutionises the way we visualise dinosaurs. Feathered dinosaurs, for instance, now number in the hundreds, whereas just thirty years ago there were none. The very reality that dinosaur colour is a genuine field of study is mind-blowing in itself! I am excited to see what weird and wonderful breakthroughs are made over the next twenty years that will surely change the face of dinosaurs again and again.

5

Who's for Dinner?

As a child, I would play for hours with my toy dinosaurs, meticulously lining them up into distinct groups and separating the carnivores from the herbivores. Soon enough, it would be dinner time for the carnivores, and the unlucky victim would usually be a big meaty sauropod like *Diplodocus*, taken down by a number of mini meat-eaters who would joyfully dine for hours. Let's be honest, if you're reading this book, then you might have done this too. I still do this, not necessarily with the toys on my bookshelf, but instead on a much bigger scale – with real dinosaur skeletons. (Clearly, all that time spent playing with dino toys did not go to waste.) However, as simple as predator-eat-prey scenarios may appear, deciphering who ate whom and what ate what in the dinosaur world is a much more difficult task than you might expect.

When looking at dinosaurs like *Tyrannosaurus*, with its enormous skull, banana-sized teeth and bone-crunching bite, it is easy for your mind to wander and begin to imagine what dramatic duels must have happened. The stereotypical fantasy dinosaur fight featured in pop culture is of a bloodthirsty *Tyrannosaurus* fighting to the death with a three-horned, giant bony-frilled *Triceratops*. Big animals, big teeth and big horns. What's not to like?

Nothing quite like these animals is alive today and this sort of imagery, of epic dino fights, easily grips our attention and fuels our excitement. The funny thing is that, even though we are yet to find direct evidence of a rex and *Triceratops* locked together in mortal combat (although a major discovery recently made in Montana, USA, may end up revealing otherwise), it does not mean it did not happen. Just as we can commonly see animals fighting and eating each other today in the natural world, there is little doubt that epic dinosaur battles such as this were a regular occurrence.

Studying dinosaur fossils is fun for so many reasons but every dinosaur palaeontologist will tell you that one of the most inextricably difficult things to understand about a long-dead dinosaur is its behaviour. Yet, the behaviour is arguably the most fascinating aspect. After all, the day-to-day life and survival of every living thing is influenced by how it behaves and the choices it makes. This is why palaeoartists — artists who dedicate their lives to bringing dinosaurs and other prehistoric organisms back to life through their art – have so much fun when it comes to reimagining ancient interactions. However, reconstructions of dinosaurs fighting or eating, as entertaining and speculative as they may first appear, are not always based on assumption.

So, how do we begin to understand what dinosaurs ate, how they ate, and which species dined on which? The first port of call is to look at the structure of a dinosaur's teeth (if it has teeth) and jaws, which can be assessed alongside those of living animals, and thus provide a basic comparison. Just look at the sharp, thick, heavily serrated steak-knife-like teeth of *Tyrannosaurus*, for example. These are clearly the teeth of a carnivore, perfectly adapted to tear through flesh and crunch through bone. Compare this tooth type with the tall and thin, forward-pointing peg-like teeth in *Diplodocus*, which are ideally suited for stripping leaves from trees.

Tooth shape and structure can provide a basic understanding of diet. Using this simple approach, palaeontologists can easily categorise a carnivorous versus herbivorous dinosaur. While these labels are useful, there is always a grey area. For instance, it is clear that *Diplodocus* was a straight-up herbivore, but it is interesting to think about *Diplodocus* munching on leaves with insects attached to them. This does not mean it actively hunted insects, but it puts an interesting spin on dino diets. The same goes for theropod dinosaurs, the group including some of the most famous predators of all, like *Tyrannosaurus*, *Velociraptor* and *Megalosaurus*. Just because the stereotypical theropods were carnivorous does not mean every theropod strictly dined on meat. Some theropods had toothless, beaked jaws, while others had teeth similar to herbivorous animals and, in reality, many theropods were omnivorous or strictly herbivorous.

Taking this a step further, nowadays in palaeontology it is practically the norm to use some flashy, high-tech computers and scanners to unlock new information about dinosaur life. Using this technology has significantly helped us to better understand how they ate. For instance, by identifying specific attachment points for muscles in the skull of a dinosaur and comparing these with living animals, palaeontologists can add muscles to the skull, jaws and neck and create accurate, detailed 3D models. These scientific models can then be used to understand things like how the animal moved its jaws, how wide its gape was and how powerful a bite it had.

Studies based on the skull of *Tyrannosaurus* found that it had a bone-shattering bite of more than 60,000 newtons, around 6.5 tonnes of force, making it the most powerful bite known for any terrestrial animal, living or extinct. It is about four times more powerful than the bite of a saltwater crocodile, which has the strongest bite force of any living animal. These findings, coupled with the massive teeth and big jaw and neck muscles, provide further evidence that *Tyrannosaurus* had the ability to chomp through solid bone. Another study that focused on the feeding habits of *Diplodocus* showed that it fed in a specific manner, by grasping tree branches with its comb-like arrangement of teeth at the front of the jaws, before then pulling its head back and raking the leaves into its mouth. Further evidence for this idea comes from the presence of microscopic pits and marks on the teeth that are indicative of tooth wear from contact with branches.

There are several other lines of evidence for feeding behaviours in dinosaurs, and one of the most obvious is in the form of bite marks on bones. Knowing which animal produced a bite mark can be very difficult to decipher. However, using some cool CSI-style detective work, by comparing the markings and their size and shape, along with looking at the teeth of dinosaurs found in the same rocks as the bitten bone, it is possible to deduce which species was the culprit. Occasionally, a tooth is left embedded in the bone of the victim, leaving zero doubts about the identity of the offender. Some rare dinosaur skeletons have even been found with healed bite marks, indicating that the animal survived the attack.

Most bite marks we find on bones are known as feeding traces, where the tooth marks were created by a predator feeding on a dead animal, like some *Triceratops* bones that were found with marks matching the teeth of a *Tyrannosaurus*. As gruesome as it sounds, similar bite marks on the frill of a *Triceratops* suggest that *Tyrannosaurus* ripped off the head before tucking into its dinner.

One sure-fire way of knowing what any animal has been eating is to look at what comes out the other end. Of course, I'm talking about poop, fossil poop. Correctly termed *coprolites*, these are fairly common as fossils and offer a direct view of ancient diets. Just as it is difficult determining who left their bite marks on bones, there are similar challenges when identifying who dung it. However, one bonus is that coprolites are often found to contain plant or animal remains and this provides a starting point. By looking at the size and shape of a given coprolite and comparing it with the size and type of animals found in the same rocks as the ancient poop, it is possible to reliably link one with the other.

The most famous example by far was an enormous, 44 centimetre (17 inch) long, chunky coprolite found in 66-million-year-old rocks in Saskatchewan, Canada, that contained crunched-up bones. By studying the animals found in the same rock formation as the coprolite, through a process of elimination researchers determined that the only predator both big enough and capable enough of producing such a big scat was a *Tyrannosaurus*.

Removing any ambiguity, some unique fossils go way beyond anything mentioned previously and give us rare and direct evidence of dinosaur diets. As absurd as this might sound, there have been some spectacular dinosaur fossils found with their last meals preserved inside their bellies. From a tiny, feathered theropod with seeds inside its gut to an armoured ankylosaur that ate leaves for its last supper, these types of fossils provide irrefutable evidence of dinosaur diet. Among the most famous of dinosaurs to be found with its last meal was the spinosaur called *Baryonyx* (my favourite dinosaur), which was discovered inside a British quarry in 1983. Contained inside its gut were the bones of a juvenile *Iguanodon*-like dinosaur and, more surprisingly, fish scales. This was the world's first confirmed fish-eating dinosaur.

Dinosaurs did not have it all their own way, as demonstrated by a marvellous fossil unearthed in China and reported in 2005. When we think of mammals from the age of dinosaurs, it is easy to imagine them as living in their shadow, scurrying underfoot and staying hidden from view, frightened that they will be next on the dinosaur menu, but one badger-sized Cretaceous mammal was found with a baby dinosaur inside its gut. Closer inspection of the dino dinner revealed that the pint-sized dismembered skeleton belonged to a small, bipedal cousin of *Triceratops* called *Psittacosaurus*. The discovery of this dinosaur-eating mammal was another world first.

Talking of world firsts, of course saving the best

for last, one of the greatest and most famous dinosaur finds of all time – and one that many would argue sits at the very top – was the extraordinary discovery of a pair of fighting dinosaurs! An actual predator versus would-be prey caught in the act.

Found in 1971 during an expedition to the Gobi Desert in southern Mongolia, this duelling dinosaur fossil captures a *Velociraptor* and *Protoceratops*, a boar-sized cousin of *Triceratops*, in remarkable detail. As preserved, the *Velociraptor* is lying on the floor on its right side and with its right arm, just below the elbow, tightly clamped in the strong beak of *Protoceratops*, which is crouched above the *Velociraptor*. The famous killing claw on the left foot of *Velociraptor* is held high in the air and in the position of the *Protoceratops'* neck, as if delivering a fatal blow to the throat.

Palaeontologists have pondered how exactly the exceptional preservation of this specific moment in time came to be. The generally accepted view is that the two were battling it out when a sand dune collapsed above the pair, burying them for all eternity. It is astonishing to think that this epic battle to the death is still to this day locked in time, captured just as they were 75 million years ago.

6

Raw Sex Appeal

Dinosaur sex is a hot topic. It's the focus of ongoing studies, conference presentations and museum displays, and palaeontologists just love to speculate on how dinosaurs did it. What about you, has dinosaur sex ever crossed your mind? Surely, you must have had that split-second thought about how two *Tyrannosaurus* actually got down or how a *Brachiosaurus* got up?

As amusing as dino sex may be, it is genuinely a fascinating area of dinosaur research because we know that dinosaurs must obviously have had sex in order to reproduce. Therefore, uncovering any potential clues preserved in fossils could shed light on this oh-so-vital behaviour.

Look to the animal kingdom today or watch some wildlife documentary, and, eventually, some aspect of sex will be discussed or observed. Unfortunately, the dinosaur fossil record has so far fired blanks. No dinosaur fossils have yet to be unearthed mid-sex, although that does not stop palaeontologists from imagining how they may have done it. Besides testing out theories by posing toy dinosaurs in all kinds of weird and wonderful positions, I have witnessed palaeontologists during conference presentations visually demonstrate (for science) the positions some dinosaurs may have used.

Of all the dinosaurs, *Stegosaurus* has become a bit of a sex icon. With large bony plates spread along its back and four spikes on its tail, it is easy to see why palaeontologists ponder how *Stegosaurus* did it. The bony plates imply that *Stegosaurus* could not have lain on its back, and the two primary hypotheses suggest that the pair either backed up into each other butt-to-butt or that the female laid down on her side and the male mounted her. For palaeontologists, working out these types of questions is both fun and frustrating at the same time.

It is not as simple as thinking what the easiest or best position may have been, because the animal kingdom shows us that sex is not all that straightforward. The mating rituals of animals differ immensely, and males and females may disagree or even fight over sex. If you were a male *Stegosaurus* trying to mount from behind, you would not want the female angrily swinging her tail spikes anywhere close to your private parts. Speaking of which, highlighting the lethal nature of the tail spikes, there is one meat-eating

Allosaurus fossil with damage to its pelvic bones that appears to match the shape of a *Stegosaurus* tail spike, suggesting that this *Allosaurus* was smashed in the crotch. Talk about a low blow.

In actuality, the closest palaeontologists have come to finding any sort of undeniable Jurassic sex scene is an incredibly rare fossil of a pair of mating froghopper insects that were caught off guard and rapidly buried together some 165 million years ago. These froghoppers were found in Jurassic rocks in northeastern China near where many dinosaurs have been discovered, so although we do not have any dinosaurs doing the deed, they were likely to have been somewhere in the vicinity as this pair of froghoppers was caught in the act.

The sad fact that we have no dinosaurs preserved in their eternal embrace is disappointing to say the least, although this has not stopped palaeontologists from searching for hard evidence. One of the main ways in which palaeontologists attempt to differentiate between males and females is through studying their 'sexy' features.

Just like the back plates of *Stegosaurus*, which have long fascinated dino fans and scientists alike, a myriad of elaborate and often downright bizarre structures have evolved in dinosaurs, from weird head crests to enormous bony frills. These types of ornamental structures probably served multiple purposes and were very likely to have been used partly for display and to attract the opposite sex.

In the animal kingdom today, physical structures such as these, along with other characteristics like size and colouration, can help to distinguish males from females and scientists call this *sexual dimorphism*. The presence or absence of antlers in deer is one of the most famous examples, with antlers being present primarily in the males of practically all deer species. Besides the obvious, this is one quick way of knowing whether you are looking at a male or a female; the males also tend to be significantly larger than the females.

However, when using such features as the basis for distinguishing sex differences in dinosaurs, palaeontologists have really struggled to show any major distinctions that can reliably separate males from females of the same species. Findings never pass by without some heated discussions. As a result, although palaeontologists would tend to agree that physical features – such as differences in head crests, body size, etc. – are somehow related to sex, the idea of sexy structures can be a tricky road to go down.

With that being said, the major finding of colour trapped inside fossil feathers may end up being the shining light of evidence that palaeontologists have long been searching for. Colour plays a very significant role in sexual display today – simply consider the luxurious plumage found in birds, like peacocks for example, and the major differences observed between some sexes. Often it is the males that have more extravagant colours, although occasionally it can be the females. So, even though palaeontologists have yet to find clear-cut evidence of colour differences between a male and female of the same extinct species, the fact that we have dinosaurs preserved with an array of colours suggests that colour played an important role in wooing the opposite sex.

At this point maybe we should shift our focus to the soft parts? Surely the easiest way to know whether a dinosaur is a male or female is to 'pull up the dinosaurs' skirts' and, well, take a look. (Apologies for the inevitable Dr Malcolm *Jurassic Park* reference.) Sadly, as amazing as the fossil record is, even with soft parts occasionally being preserved, we have yet to find a *Tyrannosaurus* penis in all its glory. Sorry to disappoint. Though, did *Tyrannosaurus* even have a penis at all?

To answer that question, palaeontologists looked at the closest living relatives of dinosaurs – the birds and the crocodylians – and inferred that the same feature(s) present in both groups would have also existed in their extinct relatives. In this case, both sexes of all living birds and crocodylians have what is known as a cloaca, which suggests that dinosaurs also had a cloaca as well. It is a single opening between the legs used for reproduction and excretion.

The penis remains tucked up inside the male's cloaca and protrudes outwards during sex, where it is inserted into the female's cloaca to transport sperm. Although all male crocodylians have a penis, most male birds lack one and exchange sperm through what is known as a 'cloacal kiss' (the touching of the male and female cloacae); but some are very well-endowed, like the Argentine lake duck, whose penis can be more than 40 centimetres (16 inches) long!

In reality, we do not need to guess whether dinosaurs had a cloaca because one has been found.

Really. It belongs to a *Psittacosaurus*, that small, bipedal dinosaur related to *Triceratops* which was mentioned in the previous chapter. The fossil was found in China and represents one of those amazing specimens that is so extraordinarily well-preserved that its skin and colour are present. Tasked with bringing this fabulous fossil to life, along with its colourful cloaca, famed palaeoartist Bob Nicholls built a 3D model of it, which has been hailed as the 'most accurate depiction of a dinosaur ever created'.

Sex is not solely about getting on down to the physical act. Before you can get that far you have to woo the opposite sex — be that by showering them with gifts, showing you are stronger than your rivals or performing some strange sex dance, which brings me to another amazing fossil and a bit of a confession. I sort of lied when I said the closest thing to fossilised dinosaur sex was two mating insects, because a phenomenal discovery of a 'dancing dinosaur sex show' was revealed in 2016.

While studying large theropod trackways exposed in Cretaceous rocks at multiple sites in Colorado, USA, palaeontologists identified many distinctly shaped scrape marks. These markings were found to match the same scrapings made by many modern ground-nesting birds which display a mating behaviour known as *lekking*. Males congregate together around breeding season to compete for the attention of onlooking females who judge them on their nest-scraping abilities, to see who can make the best nests and thus be worthy winners in the bid to entice a female. This discovery presents direct evidence that theropod dinosaurs also engaged in this type of sexual behaviour.

At the heart of all this sex talk is reproduction. We know that dinosaurs reproduced by laying eggs, rather than bearing live young like us mammals, because we have discovered thousands upon thousands of them. If you are lucky, occasionally, dinosaur eggs contain tiny developing embryos. At the most extreme are those eggs still inside the body of a dinosaur (and not as their last supper). These extremely rare fossils provide unequivocal evidence of a female dinosaur, although even then they still do not tell us much about sex and sexual behaviours, or shed any light on the possible differences between females and males.

Another approach for sexing a dinosaur is to look inside the bones for evidence of medullary bone, a specific type of temporary bone tissue found in female birds that is strictly associated with reproductive activity and used to make eggshells. In 2005, American palaeontologist Dr Mary Schweitzer and her team stunned the dino community when they reported evidence of medullary bone in a *Tyrannosaurus*, suggesting that this rex was not only female but that she had died just before, during or after laying her eggs. Later studies also found evidence for medullary bone in other dinosaurs. However, this method has received its share of criticisms and is not yet universally accepted. Besides discovering a dinosaur with eggs inside, it seems that sexing a dinosaur or learning more about their private sex lives will mostly remain an enigmatic subject, at least for the foreseeable future. Still, I hold out hope that one day we will find two dinosaurs preserved in the act of mating.

7

Family Values

Whether it is a pack of wolves feeding their hungry pups, a female crocodile patiently guarding her eggs for months or an elephant herd mourning the loss of a loved one, family life is one of the most important things for many animals. So, with all the aforementioned talk of dinosaur reproduction and eggs, naturally we begin to wonder what happened after the eggs were laid and the babies were born. Did the parents stick around or were the youngsters left to fend for themselves, and did they form close-knit groups or were they lifelong loners?

Knowing if dinosaurs were gregarious animals and good parents can tell us a lot about their life history, how they may have interacted with each other and their individual behaviours. However, as interesting as this might be, there are obvious limitations to understanding whether long-dead dinosaurs had any sort of family life at all.

In the modern world, we can step outside and watch animal groups and families interact with each other as they go about their day-to-day lives, working out who is the dominant member of a herd, whether a group sticks together year-round, or if both parents play a part in raising the young or not at all. Of course, palaeontologists do not have this luxury for prehistoric animals. The point is that something so obviously simple in the animal kingdom today can be incredibly difficult, even seemingly impossible, for palaeontologists to interpret and understand from fossils. Gathering enough evidence to reliably state that an extinct species was 'social', a 'good parent' or something similar is an almighty challenge, but it is not impossible if you know where to look.

Recall the last time you walked along a beach and left your footprints in the sand. Turning around and looking back at your footprints is looking back in time, albeit from a few minutes earlier. Recording your movements in the sand, footprints act as markers of behaviour and, just like us, dinosaurs walked on beaches and left their footprints behind too. Sometimes, when the conditions were just right, the tracks would become fossilised and the trace frozen in time for all eternity.

Dinosaur tracks are fairly common trace fossils that are found around the globe. They were made by *living*, active animals rather than representing the remains of a dead individual and offer a direct view into the world of a dinosaur. Working with trackways, it is possible to determine how fast the trackmaker was moving, how tall it was, its walking gait, what type of dinosaur left the track behind and much more. The drawback is that no dinosaur has been found dead in its tracks (yet), so palaeontologists deduce which type of dinosaur created the track by comparing its size, shape and structure with dinosaur feet. This is standard dino tracking 101.

Evidence for all sorts of dinosaurs, from small to large and carnivorous to herbivorous, has been captured in the midst of some form of apparently social situation. The most famous are those tracks found in exactly the same layer of rocks, suggesting that a large group or herd of dinosaurs might have been travelling together. Even 'mega-track sites' have been found consisting of hundreds or even thousands of dinosaur tracks left behind by numerous individuals, typically produced by herbivorous dinosaurs like sauropods.

Some sauropod multi-track sites have shown that groups were either mixed age, comprising older and younger individuals, or age-segregated, implying that juveniles or young adults may have lived in exclusive groups. In mixed-age groups, there is some indication that the smaller juveniles may have been kept in the centre of a wandering herd, while the larger individuals were on the outside. This could suggest that the adults were keeping a watchful and protective eye on the youngsters. Similarly, theropod footprints made by both large and much smaller individuals have been discovered together, hinting at possible parental care. Some other impressive theropod tracks suggest that even tyrannosaurs and *Velociraptor*-like dinosaurs may have grouped together.

As ever, palaeontologists have to be careful with their interpretations of tracks and also consider the possibility that some may have been made by animals that walked alone in the same area but at different times. Nevertheless, in situations where multiple tracks are preserved in exactly the same way, belong to the same type of dinosaur and are generally moving in the same direction and side by side, then the evidence becomes much more compelling for a group of dinosaurs taking a stroll together.

Stepping in a different direction, let's turn our attention to bones, and lots of them. Dense dinosaur bonebed assemblages representing what are known as mass mortality events offer additional support for the notion of family life and sociable behaviours. By definition, these bonebeds are essentially concentrations of bones belonging to multiple individuals contained in a single area. However, as with interpreting trackways found in the same rocks, discovering a dinosaur bonebed does not immediately suggest they all lived and died together or at the same time.

To be confident that we are dealing with a distinct family or group, palaeontologists look out for several things. Firstly, the skeletons must all be in exactly the same layers of rock. Secondly, they need to have perished due to the same circumstances over a relatively short time (minutes to days) and bear similar styles of preservation (for example, have a similar level of completeness). And, lastly, they should usually belong to the same species and be *associated* (e.g. by the overlapping of bones from different individuals). A fine example matching this description is an intriguing mini mass death of the dinosaur *Sinornithomimus*, a member of the so-called ostrich dinosaur family of theropods. A group of over twenty was collected from Inner Mongolia and was revealed to consist entirely of immature individuals, with no hatchlings or adults present, thus representing a distinct group of teen dinosaurs that was hanging out together. Unfortunately for this herd of youngsters, they became trapped inside a drying lake filled with sticky mud that led to their downfall.

One of the world's largest and most famous dino bonebeds is the 'Hilda mega-bonebed' in southern Alberta, Canada, which is made up of at least fourteen associated assemblages – hence the title of 'mega-bonebed'. Therein, the rhino-sized, horned ceratopsian *Centrosaurus* was found in an enormous mass graveyard comprising thousands of mixed-aged individuals that drowned together during a catastrophic flooding event. The mass association presents strong evidence that this dinosaur lived in huge herds and cared for its young.

Another way of looking at family life is to turn our attention back to eggs for a moment, or, more specifically, to nests. Many modern reptiles abandon their eggs after laying them, leaving the hatchlings to fend for themselves. Just think about turtles, for instance, who lay their eggs ashore and then return to the sea. Although we do not know all the ins and outs of the dinosaur world, if you ever speak to a palaeontologist and ask about dinosaur nests and parental care then one dinosaur will usually take the limelight. The herbivorous 'duck-billed' *Maiasaura*, whose name translates as 'good mother lizard' – and for good reason.

In 1977, a major discovery of a *Maiasaura* nesting ground containing eggs, embryos, hatchlings and young juveniles was uncovered at a site in Montana, USA, which became known as 'Egg Mountain'. Some of the hatchlings showed evidence of wear on their tiny teeth, indicating that they had been feeding. This suggests that the adults must have brought food back to the nests for them; incidentally, plant matter was also found around some of the nests. The discovery provided the first evidence that at least some dinosaurs gathered together and nested in colonies, perhaps at distinct times in the year, where they raised their young for an extended period. Furthermore, additional nests were found buried in rock layers one above the other, indicating that *Maiasaura* used the same nesting grounds over and over again. Does this mean that all dinosaurs were good parents? No, not at all. But it does mean that some definitely were.

As incredible as all these discoveries are, palaeontologists can still be left puzzling over them for
a lifetime. Beyond being able to confidently state
that some dinosaurs cared for their young and lived
in herds, what does this really mean for the bigger
picture? Frustrating as it is, the reality and nature
of the fossil record dictates we will sadly never
know all the finer details of these behaviours. For
example, it is impossible to say for sure whether
Maiasaura lived together year-round or split
apart; whether there was some form of herd
hierarchy; whether the herd consisted of more
females than males, or whether both parents
cared for the young. However, not to leave you
feeling discouraged, some exceptionally rare fossils
go beyond trackways, bonebeds or even nesting
grounds and provide us with an unprecedented
glimpse of dinosaur family life in action.

Among the most famous are those rare theropod
dinosaurs from China and Mongolia, found lying
atop nests of eggs. Originally, when the first of
these parrot-beaked theropods was found it was
thought to have been caught red-handed about
to dine on the eggs of another dinosaur; it was
even given the name *Oviraptor*, meaning 'egg thief'.

However, years after the discovery, the eggs were found to belong to the dinosaur itself, which was brooding them, rather like a bird. This was a seriously dedicated parent that sacrificed its life to guard its would-be offspring, shielding them from a major sandstorm that eventually buried the adult and the eggs. Telling a similar story, although a stage further, is the *Psittacosaurus* skeleton – also from China – that was found buried with at least twenty perfectly preserved juveniles by its side. The larger individual was initially thought to have been an adult, but later studies showed that it had not yet reached sexual maturity and could not have been the parent of these minisaurs. Instead, this individual appears to have been acting as a 'babysitter', left to look after the group while the adults were away.

Perhaps saving the best till last, there was also the surprising discovery of three dinosaur skeletons inside a fossilised burrow in Montana, USA. The dinosaurs all belonged to a Labrador-sized bipedal herbivore named *Oryctodromeus*. Curiously, one of the individuals was an adult and the other two were about half the adult's size and represented older juveniles. Not only does this discovery suggest that the juveniles stayed with their parent, thus indicating an extended period of parental care, but that *Oryctodromeus* excavated and lived inside burrows where it cared for its offspring, something that we would never have known were it not for this spectacular find.

8

It Came from Outer Space!

The allure and grandeur of dinosaurs is deepened only further by the mysteries surrounding their demise. How could a group of animals that had been at the very top of their game for millions upon millions of years be eradicated from the face of the Earth in an instant? Knowing that 'dinosaurs went extinct because an asteroid struck the Earth' is one of those mainstream 'facts' about dinosaurs that everybody and their dog seems to know a little bit about. While extinction may then appear to be synonymous with the word dinosaur, it is far from a dead-end topic.

The modern concept of extinction has its roots in the late 1790s and the work of a French naturalist widely recognised as the 'founding father of palae-ontology', Georges Cuvier. Through comparing fossils with living animals, Cuvier came to realise that many fossil animals could not easily be linked to, nor did they belong to, any living species. This played a fundamental role in establishing the idea of extinction. By the simplest definition, if a species dies out, it becomes extinct and is gone forever. There is no turning back the clock on extinction.

Palaeontologists deal with extinction on a daily basis; after all, the very nature of the science is focused on the study of long-dead animals and plants whose fossilised remains are the only evidence that they were ever even here. Without fossils, we would never know that species can be wiped from the face of the Earth. Through studying the fossil record, it becomes clear that extinction is a natural process, and scientists estimate that 99.9 per cent of all species that have ever existed are now extinct. Just imagine the sheer volume of species that vanished without leaving a tiny trace or subtle hint. The fossil record as we know it is nothing more than a snapshot of life through the ages.

Extinctions may happen for a multitude of different reasons. They may occur due to global catastrophic events like that which wiped out the dinosaurs, new diseases and pandemics, climate change and rising sea levels, destruction and loss of habitats, pollution, lack of food resources, overhunting, new or invasive competitors and predators, among many other things. In essence, species will become extinct when they are unable to tolerate, adapt to and survive changes in their environment, big or small.

The most catastrophic extinctions are those that

scientists term a mass extinction or biotic crisis. This is a period of rapid and global extinction of very many species over a relatively short timeframe which may take hundreds or thousands of years to play out (a blink of an eye in the geological record). In Earth's long history, there have been five big global mass extinction events, the death of the dinosaurs being the most famous and most recent. However, before we dive further into the dino-destroying asteroid, it is important to know that this is not the most catastrophic event. That title goes to a devastating mass extinction that occurred 252 million years ago before the dinosaurs were even a thing. Scientists call this the end-Permian mass extinction event or, more poignantly, 'The Great Dying', where life almost came to a halt. As many as 90 per cent of all species were annihilated, leaving room for new forms of life to emerge, including (eventually) the dinosaurs.

Many scientists argue today that we are currently in the sixth mass extinction. Although extinction is a natural process, humans have caused an astonishing surge in extinction levels, to the point that species are now going extinct at a phenomenal rate of at least one thousand times the normal speed.

Estimates suggest that a staggering 1 million animal and plant species are at risk of extinction because of human activity. Today, *we* are the asteroid.

The extinction of the dinosaurs is like the ultimate question. I guarantee that every single palaeontologist has been asked: 'Why did the dinosaurs go extinct?' If you are a palaeontologist reading this and say otherwise, then I do not believe you. It is the inevitable question that always comes up. As humans, we are both intrigued and intimidated by the power of extinction, which is perhaps why we are often so enthralled by the demise of the dinosaurs. Though it was not merely *bang!* and the dinosaurs are dead.

Over the years, more than a hundred theories have been put forward to explain the death of the dinosaurs. A common view during the late 1800s and early 1900s was that their time had just simply run its course. Later suggestions were downright bizarre. According to some, the dinosaurs went extinct due to a lack of sex drive, mass blindness from cataracts, general stupidity, constipation, disease-carrying biting insects, or because mammals ate all of their eggs. More serious theories considered that climate change or intense volcanism might be to blame. None of these held up to scientific scrutiny. Any serious hypothesis needed to be testable and justifiably plausible. Enter Alvarez and the crater of doom.

In 1980, Nobel Prize-winning physicist Professor Luis Alvarez and his team, including his geologist son Walter, proposed that the dinosaurs were shot into extinction by a giant asteroid. As far-fetched and controversial as it sounded at the time, the team was on to something. (FYI, Earth is constantly bombarded by space rocks, though practically all of them are tiny and most break up before they enter our atmosphere.) Alvarez's team had discovered rich levels of a rare metal called iridium in a thin layer of rock that marks the end of the Cretaceous and the beginning of the Palaeogene, called the Cretaceous–Palaeogene (or K–Pg) boundary. Iridium is rare on Earth but is very abundant in space rocks. They predicted that many other K–Pg sites around the world would also show high levels of iridium, hinting at a global cataclysmic event. They were correct.

It was a decade before they had their 'smoking gun' – the discovery of a colossal crater whose age coincides precisely with the extinction of the dinosaurs. Stretching approximately 177 kilometres (110 miles) across, the Chicxulub crater, as it came to be known, was found on Mexico's Yucatán Peninsula. This was the impact zone of Alvarez's asteroid, an asteroid thought to have been somewhere in the region of 10 to 16 kilometres (6 to 10 miles) wide – wider than Mount Everest is tall. Sadly, Luis died before Chicxulub was recognised as the dinosaur-destroying impact zone, but he and his team revolutionised our understanding of the dinosaur extinction. Today, palaeontologists and geologists (mostly) agree that the asteroid dealt the dinosaur death blow. This major event is known as the Cretaceous–Palaeogene (K–Pg) extinction, where as much as 75 per cent of all life was lost.

There is another question that usually goes hand in hand with the dinosaur extinction: 'Would the dinosaurs still be here if the asteroid had never hit?' In short, life would have continued to evolve and diversify just as it has over millions of years; the dinosaurs would be here in some form or another, but we would probably not be. There is also a theory that dinosaurs were on a rapid decline before the asteroid hit, apparently destined for impending doom anyway. This, however, is not the case. Dinosaurs were still very diverse at the end of the Cretaceous, with members of many groups surviving and thriving, suggesting that they were most definitely not in a terminal decline. Among the very last of the dinosaurs were the likes of *Tyrannosaurus* and *Triceratops*, who genuinely witnessed the sky fall on top of them.

One fateful day, on what may have been a lazy Sunday afternoon, 66 million years ago, disaster struck on the most immeasurably violent scale. The dinosaur-destroying asteroid crashed into the Earth at about 64,000 kilometres per hour (40,000 miles per hour), generating more than a billion times more energy than the most powerful nuclear bomb ever detonated. Any animal in the immediate vicinity was obliterated within seconds, transforming the once-rich dinosaur paradise into a silent dead world. Mega-tsunamis wreaked havoc on the seafloor and coastlines, earthquakes shattered the landscape, scalding hot debris was thrown through the air, and wildfires spread far and wide. Any dinosaur unlucky enough *not* to have been destroyed instantly had a horrible future ahead.

The full force of the impact immediately began spreading across the globe. The planet was on fire.

An enormous sun-blocking dust cloud eventually enveloped the Earth, causing a long, dark period of global cooling (an 'impact winter'). Plants needed the sunlight, plant-eaters needed the plants and meat-eaters needed the plant-eaters. The food chain collapsed. Any large land animals were toast. A bigger body means a bigger appetite and, if natural food sources are no longer available, then you are in trouble. Any animals that could not adapt quickly enough to the changing apocalyptic world met their doom.

Nothing stays on top forever. The dinosaurs had finally met their match. Having appeared a little over 230 million years ago, they went on to rule the world for an incredible 165 million years, diversifying into an enormous array of species, big and small, and reaching all corners of the globe, only to be wiped out by a huge rock from outer space. It seems almost comical that their mighty reign of supremacy would be stopped by an extra-terrestrial event. Still, dinosaurs were around for far longer than they have been gone. Though, was this truly it for the dinosaurs? Did the death of *Tyrannosaurus*, *Triceratops* and company really mark the end for the entire group or did some manage to make it through the most earth-shattering time in their history?

9

As Dead as a Dinosaur
. . . Not so Fast

Dinosaurs have long been the poster child for extinction – animals from a bygone era who had their time in the spotlight and are long gone. Most people will tell you that dinosaurs are extinct, an idea so entrenched from our early school years that it would be hard to question. Others might be bold enough to say that crocodiles or lizards are living dinosaurs, 'because they look like dinosaurs'. Occasionally, somebody might just pop up and say that 'birds are dinosaurs'; this makes me smile.

The asteroid and its wave of devastation may very well have wiped out *Tyrannosaurus* and its contemporaries, but the dinosaurs didn't go down without a fight, or should that be flight. The mass extinction paved the way for one group of dinosaurs to spread their wings and take over the world in a whole new way: the birds. Birds *are* dinosaurs. Say it with me: birds are dinosaurs.

Palaeontologists use the term *avian dinosaurs* to refer to birds and *non-avian dinosaurs* for the rest. Up until this point in the book, I have used the word 'dinosaur' to refer to the non-avian dinosaurs. However, when we hear or use the word dinosaur we should really also think about birds and not consider them as something distinct. Think of it like this: birds are dinosaurs but not all dinosaurs are birds, just as we are primates but not all primates are humans. With new documentaries and books focused on this topic in recent years, the notion of birds as living dinosaurs has become that bit more instilled in our minds.

The idea that birds are closely related to dinosaurs stems back to Charles Darwin and his revolutionary theory of evolution. When Darwin published his monumental book *On the Origin of Species* in 1859, he proposed that animals evolve over time through a process called natural selection, where those animals with features better adapted to their environment have a higher chance of surviving and reproducing, thus passing on their genes into the next generation. It's worth noting that a contemporary naturalist called Alfred Russel Wallace also came up with the idea at the same time as Darwin, although the latter is more often credited for proposing the theory.

While the significance of the theory of evolution cannot be overstated – as evolutionary biologist Theodosius Dobzhansky famously put it in 1973, 'Nothing in biology makes sense except in the

light of evolution'* – it was not Darwin or Wallace
who linked birds and dinosaurs. Instead it was one
of Darwin's staunch supporters and fellow scientists, Thomas Henry Huxley. In the same year that
Darwin had published *Origin*, the little chicken-
sized dinosaur *Compsognathus* (who we met briefly
in chapter 4) was unearthed in a Jurassic limestone
quarry in southern Germany. As if timed to perfec-
tion, just two years later, in 1861, the first skeleton
of the infamous 'first bird', *Archaeopteryx*, was discovered in one of the same limestone quarries. Just like
a bird, *Archaeopteryx* had feathers and wings, but it
also had a long, bony tail and sharp claws on its
hands – a sort of half-bird half-reptile. Studying
the skeletons of *Compsognathus* and *Archaeopteryx*,
along with other fossil (and living) reptiles and
birds, Huxley realised that their anatomy was so
similar that they must somehow be related. (This
Archaeopteryx was missing a complete head, but
Huxley predicted that *Archaeopteryx* would have jaws
with teeth, which later discoveries would prove

---

* Dobzhansky, T., 'Nothing in Biology Makes Sense Except
in the Light of Evolution', *The American Biology Teacher*
(1973), 35 (3): 125–129.

correct.) He saw *Compsognathus* as representing what a bird ancestor might have looked like and even stated: 'There is no evidence that *Compsognathus* possessed feathers; but, if it did, it would be hard indeed to say whether it should be called a reptilian bird or an avian reptile.'* The idea that birds could have evolved from a dinosaur or dinosaur-like animal was deemed quite preposterous, especially considering that Darwin's work had already ruffled more than enough feathers, and the theory sadly became lost in time.

Fast-forward a century to the 1960s and 1970s and we move our attention to US palaeontologist Professor John Ostrom and his discovery of the bird-like dinosaur, *Deinonychus*. Ostrom revived Huxley's concept that birds are connected to dinosaurs by showing that the skeleton of *Deinonychus* matched that of *Archaeopteryx* in its general anatomy. He showed that birds were members of the same generic group of dinosaurs called theropods and must have evolved from a *Deinonychus*-like ancestor. This sensation led to a renaissance in

---

*   Huxley, T. H., *American Addresses* (1877), D. Appleton & Co., New York, 43–67.

dinosaur studies. *Deinonychus* was imagined as a fast-moving, agile and intelligent animal, a far cry from the earlier view of dinosaurs as monstrous lizard-like animals that were slow and stupid. Ostrom's research transformed the way that scientists and the public thought about dinosaurs.

In chapter 3 we covered what makes a dinosaur a dinosaur, but I purposely left out one particular thing. See, the way that scientists define animals and plants today has changed dramatically. Rather than focusing solely on a combination of physical features that unite a group (or 'clade') of animals, like those that are used to group dinosaurs together, scientists also look at their evolutionary relationships (or 'phylogeny') to understand their ancestry. This approach focuses on the bigger picture of a dinosaur family tree, working out how species are related to each other and which branch of the tree they sit on.

This basically means that, in order for an animal to qualify as a dinosaur on the family tree of life, it must be descended from a common ancestor, and thus each dinosaur derived from that common ancestor will share a unique set of features. Consequently, because birds are descended from dinosaurs, they must be classified as a subgroup of dinosaurs. Not only are birds dinosaurs on the basis of their anatomy, but also through their phylogeny. Looking at it like this, birds are actually a group of reptiles and their closest living relatives are the crocodylians.

Today, palaeontologists classify birds as theropod dinosaurs within the group known as Maniraptora (maniraptorans). More specifically, the birds belong to a subgroup called Paraves, the same wider group that includes dinosaurs like *Deinonychus* and *Velociraptor*, which are among the birds' very closest relatives. This means that dinosaurs like *Velociraptor* are more closely related to a pigeon than to *Triceratops* despite the extensive separation in time.

Besides the hundreds of skeletal features linking birds and dinosaurs, other irrefutable evidence exists. Most significantly, feathers! Feathers were always thought to be unique to birds, given that they are the only living animals to have them. The discovery of many non-avian dinosaurs with feathers provided conclusive evidence that this was not the case. Even dinosaurs far removed from birds on the family tree – those not closely related – have been found with feathery filaments and some palae-ontologists have gone so far as to suggest that most, if not all, non-avian dinosaurs were likely to have had some sort of dino-fuzz. Interestingly, the current largest known dinosaur with direct evidence of feathers is the ~9 metre (30-feet) long *Yutyrannus* from China, which has feathers ranging from 15 to 20 centimetres (6 to 8 inches) long. This is a member of the tyrannosaur family and was discovered with a fluffy coating of feathers which suggests that even *Tyrannosaurus* was likely to have had some sort of feathery covering. Remarkably, dinosaur feathers have also been found trapped inside 99-million-year-old amber, including a dino-saur tail covered with feathers.

Birds inherited feathers from their non-avian dinosaur ancestors. Recent thinking suggests feathers probably evolved in non-avian dinosaurs for insulation and/or display, with flight coming along later. Speaking of display, as discussed in chapter 6, the discovery of fossil feathers with colour pigments indicates that dinosaurs could see in colour, which suggests colour must have played a key role in display. One crow-sized Cretaceous bird from China called *Confuciusornis* is known from thousands of specimens with preserved feathers and studies of their feathery fossils have found that two different 'types' of *Confuciusornis* exist, one group with long, showy tail feathers and one without them. The difference has been interpreted as strong evidence for sexual dimorphism, with the long tail feather group probably representing the males; thus, these feathers were likely to have been used for display. Furthermore, the bird–dinosaur connection, as we have learnt previously, can even be established based on direct evidence of specific behaviours, such as nesting and brooding.

The earliest known birds appeared around 165–150 million years ago, in the Jurassic Period. While non-avian dinosaurs reigned supreme during the Jurassic and Cretaceous periods, the birds were just another group of feathered theropods doing their own thing. After the mass extinction that wiped out the non-avian dinosaurs, along with flying animals like pterosaurs, there followed an explosion in bird diversity including the appearance of many modern groups of birds. Today, birds are a richly diverse group of theropods, from the tiny bee hummingbird (the world's smallest known dinosaur, living or extinct) to the largest living dinosaur, the ostrich. Just like their non-avian fore-bears, they have conquered every continent, are found everywhere on the planet, and have evolved into a wide variety of species capable of surviving in even the harshest of environments, from bone-dry deserts to freezing ice caps and tropical forests.

A sparrow, emu, penguin and pelican are just as much of a dinosaur as a *Stegosaurus*, *Triceratops*, *Tyrannosaurus* or *Brachiosaurus*. Dinosaurs are the most successful group of land vertebrates alive today. The next time you see a bird, remember that you are looking at a living dinosaur whose origins extend back to the Jurassic. With more than 10,000 living bird species, the reign of the dinosaurs is far from over.

10

We've Only Scratched the Surface

Every time I finish a talk on dinosaurs, I prepare for a flurry of impending questions. Chatting about dinosaurs with the public is thoroughly enjoyable, though you can never quite prepare yourself for an open Q&A, so you have to go in expecting the unexpected. After all, this is the audience's opportunity to grill a palaeontologist. Like some wondrous prehistoric being with superpowers, the palaeontologist is often expected to know everything about dinosaurs and ancient life. But not having the answer can actually have far greater impact – the unknown breeds curiosity and curiosity leads to discovery.

One of the most curious questions I get asked is: 'Do you think we will ever find all of the dinosaurs?' On the surface, it seems like such a simple question, yet it has far wider implications. In theory, all the dinosaurs that could ever have been preserved as fossils are already fossilised. This means that there *is* a finite number of dinosaur fossils that could be found. Long before humans began studying them, countless dinosaur fossils had already been lost and destroyed by natural processes over millions of years. Couple this with all the talk of new dinosaur discoveries here or new species there and it might seem like we could one day have dug them all up. The reality is that we will never run out of dinosaur fossils, and we will never find them all either.

We are currently living in a golden age of dinosaur discovery. On average a new species of dinosaur is discovered every other week. Every other week! Up to this point, in almost 200 years of study, palaeontologists have identified around 1,500 different species of dinosaur. However, sometimes new and more complete finds can help to reveal that previously named dinosaurs are actually the same as newly discovered ones and not different species at all. They might be juveniles and adults of the same animal, for example. Therefore, the total number of known dinosaur species fluctuates depending on the discovery of more dinosaurs, new research, and new information.

With so many new dinosaurs being discovered it is hard to keep track of them all. If you think really hard you might be able to get to fifty or more, but 1,500? I doubt it. (No matter how much of a hardcore dinosaur fan you are.) There are far more dinosaurs out there than your standard *Stegosaurus* or typical *Triceratops*, and when I ask people to name their favourite dinosaur and they tell me some really obscure name I have to ask them what it is! This highlights the immense quantity of new discoveries being made right now and the hard work being undertaken by dedicated palaeontologists and fossil hunters. For instance, have you ever heard of the winged wonder called *Yi qi* or the long-necked *Leinkupal*? Or what about the 'nose king' *Rhinorex* or tyrannosaur *Thanatotheristes*? (No, I can't pronounce that last one either.) These are only a few of the many dinosaurs discovered and named over the last few years.

With new dinosaurs emerging thick and fast, we must remember that fossilisation is an incredibly rare event. For a dinosaur to be preserved as a fossil, it must have died in very specific conditions that suited its potential preservation, plus be in a situation that meant no other dinosaur could eat it, or the environment destroy it. We cannot, therefore, expect that individuals of every single dinosaur species that ever lived could have become fossilised. The same can be said for all the animals alive on the planet today. There is zero chance that every single species will be preserved in the fossil record because not all living species exist in environments that are conducive to fossilisation. It is therefore important to consider that the fossil record is biased in that dinosaurs found as fossils were only preserved because they were in the right place at the right time (well, actually, many were in the *wrong* place at the *wrong* time and died as a result, but you get the point). Plus, we humans also have to be looking in the right places and in the correct types of rocks to even stand a chance of finding a dinosaur.

One of the other major factors is that the fossil record only preserves certain slices of geological time, and many rocks have already long since eroded away, meaning that we will never get anywhere close to a complete picture of the whole story of prehistoric life, let alone the dinosaur world. Even Charles Darwin had this view of the ancient world, stating: 'I look at the natural geological record as a history of the world imperfectly kept and written in a changing dialect.'*

---

* Darwin, C., On the *Origin of Species by Means of Natural Selection, or the Preservation of Favoured Races in the Struggle for Life* (1859), John Murray, London.

Just think about the reign of the dinosaurs from ~230 to 66 million years ago and the ~1,500 different species identified. Now compare that figure with the more than 10,000 living species of dinosaurs (birds) today and you will quickly realise that there must have been an unimaginable number of dinosaur species spread over many different intervals of time during their roughly 165-million-year reign. This makes you wonder what other weird, wonderful, and bizarre dinos were once here, but whose stories will never be told. It's quite sad to think about these ancient animals that will be forever lost in time.

Realising that we will never find all of the dinosaurs is horribly frustrating, though it does leave room for speculation. Just because we have yet to find a certain dinosaur does not mean it did not exist. I am not talking about a fantasy world with a dinosaur with two heads (although one dinosaur-aged reptile was found with two heads, the result of an abnormality) or something similar, but perhaps we are still waiting to find the biggest of the big or maybe that elusive pair of mating dinosaurs. By studying the dinosaur fossils that we do have, it allows palaeontologists to make educated guesses as to what might be found in the future. Many predictions have already come true, with perhaps one of the most famous being the idea that one day we would discover feathered dinosaurs. Without feathered dinosaur fossils, the dino–bird connection would always have been based on similarities in the skeleton, but the presence of feathers was the ultimate clincher showing the link to be fact, not fiction.

When it comes to deciphering who was the smallest or largest or who had the biggest bite, these are cool topics to study, but they are not necessarily at the top of the research list for palaeontologists. Some of the most important research areas to explore focus on the bigger pictures of evolution, particularly the earliest known dinosaurs and their ancestors. Although the earliest dinosaur fossils date from just a little over 230 million years, these do not represent the earliest dinosaurs to evolve. Instead, considering that the dinosaur fossils from this time already represent various different types, it means dinosaurs must have evolved some millions of years before. To support this assertion, some possible dinosaur tracks have been recorded from rocks 240-plus million years old. As no bones have been found, we cannot say whether these tracks were made by an actual dinosaur or a close dinosaur cousin.

The problem is further compounded by the fact that many fossil reptiles from earlier Triassic rocks – where we think the earliest dinosaur might be – look like dinosaurs but do not quite fit the 'what is a dinosaur?' definition. Some of these early dinosaur-looking reptiles, like the 243-million-year-old *Nyasasaurus* from Tanzania, East Africa, are known from just a few isolated bones. Not enough to say whether they are from a bona fide dinosaur. Naturally, for animals like *Nyasasaurus* we need a more complete skeleton to reveal its true identity. Much easier said than done.

Another approach to moving the science forwards is to go backwards by looking at 'de-extinction', the process of (apparently) bringing extinct species back to life. As far-fetched and *Jurassic Park*-like as this may sound, some scientists have made enormous strides. Firstly, I must point out that finding DNA in multimillion-year-old dinosaur fossils, including in amber, is to this day impossible because DNA breaks down swiftly over time. Even if we did find dino DNA or blood cells (which have been identified), they would still have been altered and damaged through the process of fossilisation and thus could not be used to recreate a dinosaur. However, some groups of scientists have taken an alternative approach by 'reverse engineering' chicken embryos, tweaking the DNA so that their beaks become modified reptile-like dino jaws. While others have looked at growing long tails on chicken embryos as well. This process would essentially mean creating an extinct dinosaur lookalike from a chicken. In case you were wondering, none of these 'chickenosaurus' have ever hatched. At least, that's what they tell us.

A major research area in de-extinction focuses on the resurrection of the Ice Age woolly mammoth. This is a more realistic possibility given that numerous, often exceptional, specimens with soft parts are discovered in the Siberian permafrost. However, if this were to be successful, the resulting creation would never be an actual mammoth, but some oddball hybrid crossed with a modern elephant. In this situation, I think it would be better to invest time, money and research into saving what we have, those animals that are at real risk of extinction, like rhinos, orangutans and elephants, to name but a few. Sticking with the

Jurassic Park theme, although a fictional character, Dr Malcolm's famous line, '*Your scientists were so preoccupied with whether or not they could, they didn't stop to think if they should*', holds very true in this situation. Perhaps the compromise could be to look at bringing back species that humans made extinct, like the passenger pigeon or Tasmanian tiger.

The beauty of science is that it is constantly changing and we are always learning. Palaeontology has come a tremendously long way from the early dino discovery days of the nineteenth century. Even within the last thirty years, the science has progressed so radically with new discoveries that our general perception has changed from 'dinosaurs are extinct' to 'birds are related to dinosaurs' to 'birds might be the descendants of dinosaurs' and finally that 'birds *are* dinosaurs', which marks a major revelation in the field. At this pace, who knows what we may uncover about the incredible dinosaur world in the next twenty, fifty or hundred years? Realising that we will never have all the pieces of that gigantic dinosaur puzzle is what continues to spark our childlike fascination and natural curiosity with everything dinosaur. There is one thing for sure, our love of dinosaurs will never go extinct.

Acknowledgements

Perhaps the most rewarding aspect of palaeontology is helping others to understand the incredible nature of this awesome science, to think more about the world around them and appreciate their place in time and space.

This is why writing a book is a lot of fun, but also a lot of hard work. This book is a culmination of many years spent studying and writing about dinosaurs and other fossils, working with some of the best minds in the field, and keeping up to date with the latest and greatest finds. Reading through various scientific studies and academic tomes helped bring this book together, with a notable mention of the excellent book, *Dinosaurs: How they lived and evolved*, written by two of my friends and fellow palaeontologists, Dr Darren Naish and Dr Paul Barrett. Admittedly, it is difficult to list the many palaeontologists and dino enthusiasts who have helped me in my career, or who have indirectly helped with the creation of this book, but if you are reading this acknowledgement please know that I appreciate everything you have done for me.

It is impossible for me to write a dedicated dinosaur book without acknowledging my good friends and colleagues at the Wyoming Dinosaur Center, who gave an eighteen-year-old dino-mad kid from Doncaster a chance at making his dream become a reality. It was ultimately my first trip to Wyoming in 2008 that formed the backbone of my career.

Above all, thank you to my wonderful mum, who encouraged me at every step of the way, doing so with love and support throughout my entire life. To Natalie, thanks for everything you have done and continue to do for me, including reading the first version of this book – you have helped in more ways than you will ever realise. To my good friend and fellow palaeo dude, Jason Sherburn, thanks for reviewing the first version of this book and providing some excellent comments to make it better. And finally, thanks to my nan, brother, sister, dad, brother-in-law and niece and nephew for your constant support.

One final note, from me to you, dear reader. Never let anybody tell you that you cannot achieve something. My journey to becoming a palaeontologist certainly did not follow a traditional path. I always knew what I wanted to do but figuring out how to get there was the hardest of all. In school I was an average student at best and struggled for most of my school years. Due to poor grades, I was not allowed to do A-level science and I never went on to do an undergraduate degree. On multiple occasions I was told that I would never become a palaeontologist – that I simply 'wasn't good enough'. The point is, if I had listened to those people, I would not be the person who I am today. It is important to find your own way in life and discover what works for *you*. Never let anybody discourage you from following your dreams.